The Avocado Pit Grower's Indoor How-to Book

With fourteen line illustrations by
Timothy Perper

A William Cole Book

WALKER AND COMPANY New York

The
Avocado
Pit
Grower's
Indoor
How-to
Book

by HAZEL PERPER

CONTENTS

PREFACE

This is a rebellious book—an act of quiet but desperate necessity.

Anyone—mailman, neighbor, delivery man, child, doctor, lawyer, Indian chief—who enters my living room for the first time inevitably says, "It's beautiful! What is it?" It's an indoor avocado plant. It is ceiling high, with a spread of six feet. Some of its glossy, dark green leaves are fifteen inches long. Two and a half years ago it was the pit left over from a delicious lunch.

It is this handsome green despot that has created the social condition in my household which I now hope to change.

Writing a how-to book on indoor avocado growing is a self-defensive action intended to liberate me from a number of small harassments —from the midnight telephone call ("We've just done the dishes and we're ready to toothpick the avocado, but I forgot whether you said to *wash* it or not") to the unexpected coffee break with an upstairs tenant ("You don't know me, but I'm from 14-H and the doorman told me about you: so *you're* the avocado woman").

Bad enough to be a sort of botanical midwife, called out of the night at odd hours on pointers on toothpicking a pit. Worse still to hear oneself referred to as an Avocado Woman. But this situation becomes nearly intolerable when everybody who comes to our home talks about nothing but avocados. The avocado is instantly the first and exclusive subject of conversation whenever a group gathers in my living room. If there is a newcomer present, I know that I will have to spend a good part of the evening talking about the tree and its care and growth. I can predict to a nicety the way the conversation will go, rating the probable incidence of comments, questions and so forth about as follows:

> "That's an *avocado*?!"   A certainty.
> "How did you do it?"   Excellent chance.
> "How long did it take?" Very good chance.
> "I once tried one but it grew straight up and didn't have many leaves. What did I do wrong?"   Fair chance.
> "What kind of soil?"   Only a possibility.
> "Will it grow fruit?"   Absolute certainty.

And when answers are offered, or when I try to outline an overall theory:

> "You have to cut back the first stem."
> "*What?* The first *stem,* you mean cut it off?"
> "And keep cutting back."
> "Really?" An expression of surprise and a dubious glance, followed by a walk to the tree and a close look into the foliage. "Hmmmm."
> "It's a tropical tree and has to be kept warm."
> "You don't say."
> "Never use cold water."
> "Is *that* it? I never knew that." And, at times, an envious note is sounded. "You must have a green thumb."
> A vehement denial on my part, but a chance, at last, to change the subject.

After months of being tyrannized by my conversation piece, I decided life was far too full of a number of other things to give so much time to the repetition of the same formula. *Voilà: "The Avocado Pit Grower's Indoor How-To Book."*

Though I hope the emergency phone calls are a thing of the past, I plan to present copies of this book to friends who might otherwise feel abandoned in mid-plant. For, all rebelliousness aside, I do not forget how attached one can become to something green that one has grown, by oneself, indoors, and all the way up from a mere seed. It's a peculiarly satisfying experience.

10

I ❃ THE AVOCADOS AVAILABLE, AND THEIR SOURCES

Many people have tried, at one time or another, to grow an avocado tree indoors. The instructions they follow are usually word-of-mouth, and often something important is left out. They start the pit upside-down in a glass of water; they get discouraged when nothing happens *immediately*; they don't know how to plant the pit; they don't know how when or why to cut back the stem.

Avocado-growing is really very simple. It takes little of your time. And you grow a stunning plant—an indoor tree for which florists could charge an exorbitant price.

Every stage of the growth is fascinating to watch: from first roots and first shoots to the first tiny leaves, held tight and closed at the tip of the

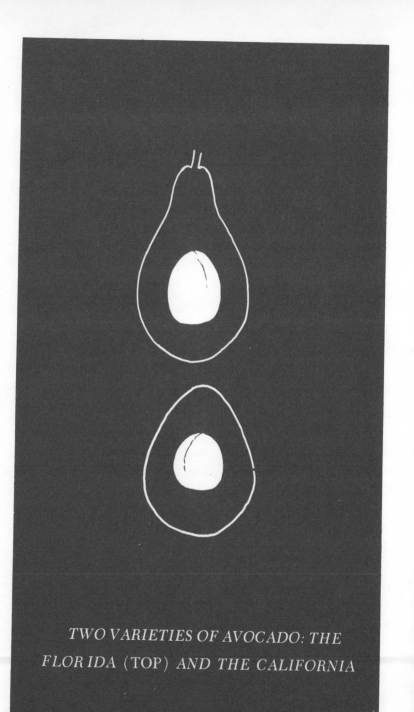

*TWO VARIETIES OF AVOCADO: THE
FLORIDA (TOP) AND THE CALIFORNIA*

stalk, to the gradual leafing out of the full plant. The avocado is hardy and cooperative. You can control its size—if you don't want a tree, you can keep it plant-size. It will last indefinitely. I've had many of them for years and have only lost them by giving them away.

Many different kinds of avocados are sold in American markets. As a result of extensive hybridization they vary widely in shape, weight, color and flavor. Although some fruit comes to the United States from Puerto Rico, Santo Domingo and the West Indies, most of what we get —and for simplicity's sake I'll stick to this—comes from Florida and California.

FLORIDA

This is the larger fruit and, in general, the best and easiest to work with. It may vary in weight from ten ounces to a jumbo forty ounces. The color may be light green or fairly dark with a brownish cast and an almost-smooth texture. Or some—generally the larger ones—may be a dusky purple, with a pebbled and roughly mottled surface. The flesh is smooth and soft, high in oil content. The pit is not as large, in proportion to the size of the fruit, as that of the California variety. The pit of the Florida fruit is more likely to "take" and it usually grows more rapidly than the other.

13

CALIFORNIA

The California fruit is quite often stamped with California trade names on the skin, which is green and smooth, although occasionally slightly pebbled. It weighs six to sixteen ounces. Its flesh is firmer and more watery than that of the Florida fruit, and it has a lower oil content. The pit is large in proportion to the size of the fruit. California avocados are much harder to start; they grow more slowly than the Florida variety and they produce smaller—although charming—plants.

HOW TO JUDGE RIPENESS OF FRUIT

The easiest and fastest way to grow an avocado plant is to start with the seed from a ripe fruit. (It's better eating, besides.) There is one sure-fire test for ripeness, even though it is less than popular with shopkeepers—the judicious application of a furtive thumb. Press at the stemmed end of the fruit. If it gives, even a bit, the fruit is fairly ripe. If it's soft, the fruit is fully ripe. But if it stonily resists all pressure (and is the only one to be had), take it and let it ripen at home. Don't keep it in the refrigerator. Put the fruit in a brown paper bag and set it aside. But keep an eye (and a thumb) on it: the fruit can ripen within a day or so.

If your shopkeeper tells you (as I've sometimes been told) that he prefers to carry the California avocado because it doesn't turn black when it is opened, like the Florida, I'd like to make a point: the flesh of *any* variety of avocado darkens when exposed to air for any length of time. But there is a remedy for this. If you plan to use the avocado in a salad very soon after opening it, a bit of lemon in the dressing will prevent the avocado from darkening and keep your salad green. If you are preparing a salad well before serving time, or are using a dressing without lemon, you can squeeze a little lemon juice on the avocado flesh to preserve its attractive color. Store any cut-open avocado with a piece of waxed paper over it. If there are leftovers, sometime you might try mashing a chunk of avocado into a hot, clear soup just before serving—it adds a delicious flavor.

THE BASE OF THE PIT IN WATER

2 ❧ STARTING THE SEED

You'll need: A clean glass not less than five inches high and with a wide opening at the top.

Wooden toothpicks. The rounded ones are stronger and easier to work with, especially if the seed is not very ripe.

❧ IDENTIFYING THE BASE OF THE PIT

The roots of the avocado plant grow out and down from the base of its big seed. (Once a plant has started to grow, that useless discard, the "pit," can accurately be called a "seed.") But if you've never before looked at an avocado pit closely, you may be wondering which end is down. It is the place where a folded-in dimple can be seen at the bottom of the pit. Many pits are tapered up and away from a broad and clearly defined base (the fruit is not called the alligator "pear" for nothing). But some are quite round, or oval, with a flattened end. Look for the dimpled bottom if you're not quite sure which end of an avocado pit is up.

❦ PREPARING THE SEED

Once removed from the fruit, the pit may be a slippery object, wrinkled and pale, undivided and quite featureless, without much in the way of a coat. Or it may be roughly mottled, with a brown and distinctive papery coat. It may also be partially split, showing the first tendrils of growth clustered inside it, and the first roots putting out from the base. But no matter what its stage of development, it should be washed.

> Right here and now an important fact should be firmly fixed in mind. *At no time should cold water be used on the avocado.* The plant is tropical in origin and will not take kindly to cold temperatures of any kind.

❦ WASHING THE PIT

Rinse the pit in tepid water, removing as much skin as comes free easily. Don't dig into it. Be a little careful when handling a widely split-open pit. The two halves should not be entirely separated: if they are, a vital connective thread between them will be broken, and the seed's fertility may be impaired. But don't be too concerned about the way you handle the seed. By and large, it's tough and hardy, like the tree itself.

PREPARING THE PIT FOR YOUR GLASS

Dry the pit and gently wipe it off, peeling away any stray shreds of skin that may still adhere to it. Put it aside. Fill your glass with warmish water. Take up the pit again, and get a good grip on it. Then, about a third of the way up from the base, force half the length of four toothpicks into the pit. Place them at regular intervals, making a framework to support the pit across the top of the glass. (See illustration.) If the pit is very hard, or if you're not much of a construction engineer, use more toothpicks; they can't do any damage. Now place the toothpicked pit across the top of the glass, allowing about half an inch of water to cover the pit's base.

WHERE TO KEEP YOUR GLASS

The pit will now stay in water until it puts down roots. Place your glass in the warmest spot you can find, and one that is not subject to sudden icy

19

drafts. And keep it out of strong light, natural or artificial. In dimness or shade the downward development of the roots will not be distracted by light from above. Your kitchen may be a very good place, warm and with water at hand, and perhaps a closet available to ensure good, dark shadows. Maintain a constant level of water in the glass at all times. The base of the pit should always be immersed. Be sure the pit is not exposed to gas fumes, which are pure poison for most plants, including the avocado.

Watch for the first signs of activity to appear.

❧ FIRST SIGNS OF GROWTH

It may only be a matter of a few days before the first roots appear, and when they do, your pit is now a full-fledged seed. But it may take longer to start rooting, so keep the seed where it is, and watch and wait. As long as the water in the glass stays clear, the seed is still sound. If you really get impatient, you can take the braced avocado seed from the water and sneak a look underneath it. You may see the first signs of life, a slender rootlet or two uncurling—or there may be no change in the seed's appearance at all. Be patient. Put it back in the water, back in the shade, and give it more time. I've had pits in water for as long as four

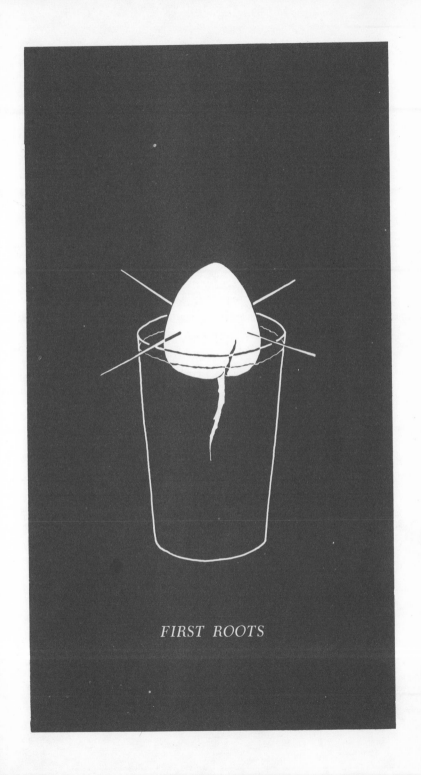

FIRST ROOTS

weeks before they began to germinate. Remember that as long as the water remains clear your avocado seed is still healthy. Signs of decay will make themselves known soon enough. The water will begin to thicken and become cloudy, and a decaying pit has a rather unpleasant odor. Dump that pit and start another. But if the pit was not sterile to begin with, and if it is kept warm, wet and dimly lit, sooner or later it will certainly germinate.

SECOND SIGNS OF GROWTH

There are other signs of life to watch for, aside from the appearance of the first roots.

The seed will begin to split. If you started with a seed that had already clearly opened, you can be quite sure it's going to go on and become a plant. But if your seed is one of the tightly closed ones, its splitting is the next thing to notice. For me, this inevitable development is one of the most exciting parts of the whole process.

THE SEED BEGINNING TO SPLIT

SPLITTING OF THE SEED

Bit by bit—and it can happen in a few hours or take as long as several weeks or more—the big seed eventually separates. There, lying in the center of it, is the first fine, pale-green tendril, ready to shoot out and up into the air.

In some seeds there may be as many as three or four of these latent shoots in sight. They can start to grow rapidly, but one of them always takes the lead and will become the main stem or trunk of the plant. This main stem grows up and up, with the first tiny leaves held tight and closed at the tip of the green stalk. Or, in others, the reddish, freckled stalk may be only a few inches high when the first bronze-colored, shiny leaves begin to uncurl. Obviously these are different varieties of the avocado. The concealed differences, micro-miniaturized in the obscure and unknown pit, are only revealed when the seed begins to germinate and grow. Your plant is beginning to tell you more about its own, *specific* character.

THE FIRST SHOOT

23

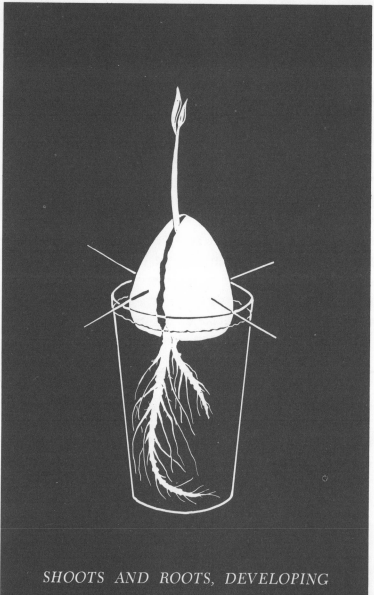

SHOOTS AND ROOTS, DEVELOPING
TOGETHER

A good growth of roots should develop with the new greenery. These roots are put out in several ways. They can be generous—falling thick and fast—and full, or they can be somewhat sparse—growing slowly—with a single root, thick, round and noduled. When both roots and stem develop at the same time, the seed has entered a state of good, active growth.

But an attempt must now be made to balance the development between the roots and the stem, and preference must be given to the roots. This is done by cutting back the stem, which serves to give the roots a better chance to put down while at the same time it checks the stem's too rapid growth.

25

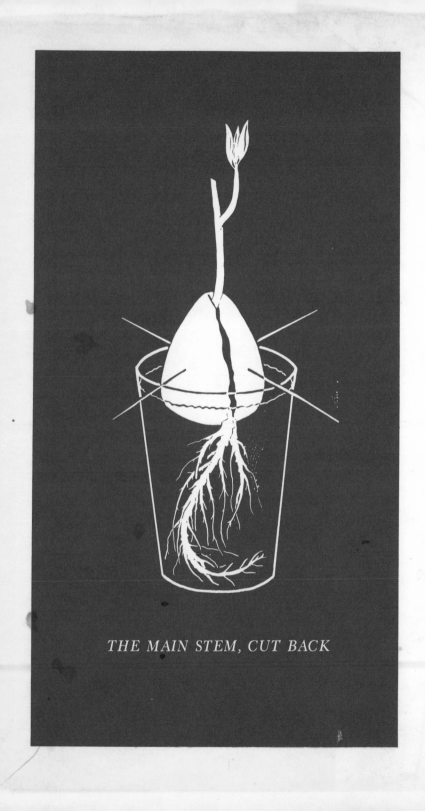

THE MAIN STEM, CUT BACK

3 ❧ CUTTING BACK WHILE THE SEED IS IN WATER

You must try to be firm about cutting back an avocado's foliage. In their natural environment some trees grow straight up to forty feet. Others reach a more modest height of about twenty feet but are fuller and bushier. To achieve a many-branched and fully fruited tree, outdoor horticulturists control the tree's shape by cutting it back. The same rule applies to an indoor plant, but with a few differences. If you want a lush, full, leafy but reasonably sized house plant, you must start cutting back at the very start. This cutting back is continued as the plant grows.

A seed may have several shoots growing up from its split. They look like smaller stems. Don't cut back on them, just leave them alone. They will flourish, or not, when the seed is put into soil.

There is no mistaking the role of the main stem, however. It takes over at once, with great conviction. This slender column may already be tipped with leaves, opened or closed. Allow it to grow to a height of six or seven inches. Then, using a scissors, cut off the stem at a point mid-way between its top and bottom. Don't cut back *too* far. The remaining stem should be at least three inches high, but no shorter than that. By way of reassurance, let me explain that if you don't cut back on the stem here and now, that stem will continue growing, on and on, producing leaves one by one. But they may also fall off, one by one, and the plant, left to its own devices (in soil or water), will finally look like a telephone pole with only a few isolated leaves dangling from its tip.

The cut-off stem's development is now frozen. It may take another week or longer to start up again. Don't be impatient. The seed has to re-group its resources, and this takes time. When the new shoot does appear it will grow out from somewhere along that three-inch stalk. Its growth will be slower, as will be the development of the leaves.

*TIME FOR THE GROWTH OF
ROOTS AND STEMS*

Roots. It is difficult to answer the question of how much time is needed for upward and downward growth to begin. The first roots may appear in water in only a few days, or they may take four weeks or more to start growing.

Stem. The stem may appear in very little time, for in some very ripe seeds and in some varieties where the seed is already split open the stem can be seen before you lift the pit from the fruit. With other varieties, and depending on the ripeness of the fruit, it may be days or even several weeks before the stem emerges.

Even if you leave the seed in water for as long as six weeks you won't jeopardize the plant's future growth. Don't leave it too much longer, however. The seed can go bad if it's not put into earth within a reasonable time. The roots should be given space in earth before they look as if they might burst your glass.

WHEN TO POT

You will be able to judge the time for potting in several ways.

1. A good, thick fall of roots is needed. They should be fairly dense and full, and long enough almost to reach the bottom of the glass. (But if the stem is growing hand over fist, even after you've cut it back, and the roots remain only

29

fairly thick, plant the seed immediately. It's a different variety of avocado from the kind that puts out the slender but quicker-growing roots.)

2. You should plan to plant the seed about two weeks after you cut back the stem. It can be left longer, but don't postpone the transfer for more than three weeks.

It is desirable to have a good root system well established before you plant. However, no exquisite timing is necessary. What is needed is to get the plant into earth fairly soon so that it can start producing its own nourishment from the light and soil, rather than from its seed.

SIDE VIEW OF THE
POTTED SEED IN SOIL

4 ❀ PLANTING THE SEED IN SOIL

You'll need: 1. Flower pot. 2. Dish to go under pot. 3. Broken crockery, or an expendable flower pot, broken. 4. Soil. 5. Plant food. 6. Dowel or stick.

❀   *NOTES:*

      1. Flower pot. The diameter at the top should be 8½ or 10½ inches.

      The flower pot you use is of prime importance to your avocado's welfare. The avocado's roots must have free soil drainage at all times. Avoid so-called "self-watering" pots: they do not give a sufficient flow of water. Get the familiar kind of inexpensive terra-cotta pot that sweats. Avoid painted pots: they tend to hold moisture inside them. If you have the space for it, start with the 10½-inch pot. You can by-pass a later transplanting step if you do. But the plant will do well in the 8½-inch pot if you prefer to start with that size.

2. *Dish to go under pot.* Don't use a terra-cotta dish. It oozes and will water-stain the surface of whatever it is placed on. Avoid metal pans. Unless they're of high-grade stainless steel or aluminum, they rust, look messy and finally fall apart. A ten-inch pyrex glass pie plate is a good vessel for holding water and can be cleaned easily.

3. *Broken crockery or an expendable or broken flower pot.* These are to be used, broken-up, placed around and over the drainage hole in the bottom of the pot before you add the soil. This prevents too dense packing of the root ball. A small flower pot, 3½ inches in diameter at the top, can be used, or an equal amount from a broken cup or two. Fold a sheet of newspaper over the pot or cups, and using a hammer, break the crockery into chunks. Don't crush or break into too small pieces. The 3½-inch pot, broken into six or seven pieces, is about right. What one is after is a set of curved shards to be placed over the drainage hole to keep the soil there loose. Don't use pebbles.

4. *Soil.* The avocado can grow in almost any kind of soil. It will do better, of course, in

a rich mixture. Five-pound sacks of sterilized, humus-enriched soil are sold in all gardening supply stores. You will need more than one bag if you are using the 10½-inch pot. Or if you're in the country and you see a good-looking patch of rather rich soil there, dig up a bagful and use that.

5. *Plant food.* Any commercially sold kind will do. Most of them have balanced and required amounts of nitrogen, potash and so on.

6. *Dowel or stick.* One at least three feet tall. Another up to eight feet, if you're planning to grow a tree.

The plant's stem should have the support of the dowel as soon as it reaches a height of sixteen or seventeen inches. If your plant turns out to be tall and thin-stalked, the dowel will have to be put alongside the stem a few weeks after planting the seed. Place it an inch or two away from the seed and shove it all the way to the bottom of the pot. But do it carefully, so as not to injure too many roots. Use green string to fasten stalk to the support.

You can buy these sticks at most supply stores, painted and fairly expensive. Or you can buy a ten- or twelve-foot length of ⅜-inch doweling at any lumber yard for much less and cut it into the lengths you need. You can color it yourself. Get some green casein or poster color paint (both are soluble in water). Brush on the paint in

33

long strokes and then wipe off with a dry rag before the paint dries. You now have a pickled stick with a matte finish and you can cut off a three- or four-foot length for the first planting and have a piece left for later needs. Don't use oil paints or chemical stains.

POTTING

The avocado seed should be given shallow planting. Plan to leave at least half of its upper mass exposed when it is potted.

Arrange the broken crockery at the bottom of the pot. Shake a sufficient amount of soil across the shards to cover them. Don't tamp down the soil.

Now take a look at your avocado and decide how much space in the soil and the pot you are likely to need for it. Keep in mind that the soil should be lightly contained, and that it will settle of its own weight when it is watered. But remember, too, that you want to leave more than half of the seed exposed at the top, so don't be too concerned with how high the soil comes to the top of the pot; you can fill it quite full. The seed needs room at the top to put out whatever other stems may be still lurking inside it. Excess earth will be washed away and down from the seed when you water the potted plant.

THE PLANT GROWING IN SOIL

Take the seed out of the water and remove the toothpicks. They can be broken off close to the seed if they don't easily pull free.

Plant the seed. This may take a bit of jockeying around, trying to get the seed and its roots into position. Just take care, however, not to pull on the plant too strenuously; you will break off roots if you do.

Now put the potted plant into place and, using the same water the seed was in before, water from above, dribbling over the split in the seed. Then fill the plate with *tepid* water, and pour more on the soil from above. The water from the top, combining with that from below, will lift water into the soil from the plate by capillary action. Wait a few minutes, and then water until the soil is fairly well soaked. You can add more soil if needed. Leave an inch of water standing in the plate. Remember, *use tepid water*.

Your avocado is now planted and ready to be put where it will best flourish.

WHERE AND HOW TO KEEP THE PLANT

Now that the young main stem has been cut back, the plant's aerial growth is at a temporary standstill. This can be an uneasy period, while you wait and watch; it may be weeks before the stem starts up again. However, a new but shorter stalk will finally put out from the truncated stem. The new stalk will appear sooner if the plant is placed in strong light.

From now on, your plant should have as much light as you can give it.

The most convenient source is sunlight. If there's a spot in your house that enjoys several hours of sunlight a day, this is the best place for your plant. But the plant will respond just as effectively to artificial light, provided it's strong enough and is supplied often enough.

The direct light from two 100-watt bulbs for a few hours a day or more is adequate. Don't burn the leaves by letting them touch the light bulbs.

To ensure an even, constant distribution of light, change the plant's position from time to time, turning it in the light.

Several hours of light from a frosted white fluorescent tube is excellent, second only to sunlight. This light source is safer and is superior to the ultraviolet fluorescents that are having such a vogue among indoor gardeners. A plant is much more likely to be burned to a shriveled crisp than to benefit from the intensity of ultraviolet.

WATER (NEVER COLD, ALWAYS TEPID, AND EVEN WARMISH)

Water once a day or more if necessary. You can water at any time of the day, but performing the task at the same time each day has the advantage of becoming a habit.

The most desirable objective is to keep the plate underneath the pot always wet. The plant thrives if it is kept good and moist and if it has full, free drainage through its root system.

However, an occasional short lapse from watering won't seriously harm the plant. If you're going to be away for two or three days, just fill the plate with water and don't worry about your avocado. On your return, water the dried soil from the top first, and then fill the plate with water until the soil is well soaked. The drooping leaves that result from a drying-out will pick up at once.

Try not to parch your plant too often. And be watchful if you live in a temperate climate where the indoor air is very dry in winter. A prolonged drought will destroy the plant, just as too many dehydrations will injure it.

A long-nozzled watering can is useful. It allows you to fill the plate underneath the pot without spilling, and every once in a while it can be used to direct a stream of water over the leaves while the plant is still small.

TEMPERATURES

Cold. Anything under 36 degrees Fahrenheit is risky for an avocado. Some species withstand a little frost, but if extreme chilling can be avoided, it should be. Guard your avocado against wintery blasts from open windows.

Heat. The plant will thrive in any degree of summer's temperatures. During the summer months the plant can be put outdoors. If you have a terraced apartment, the plant can be kept there all during the summer.

Your plant can be kept on top of a radiator. But be very certain that there is a heavy or asbestos cover over the hot pipes. If heat becomes too extreme or too direct from below, the plant can be cooked, which is a bad thing. I keep a number of plants atop a radiator that has an asbestos protective layer over its surface. Drafts of hot air stir the plant's leaves, and they flourish and grow, perhaps because there is such a tropical atmosphere above, as well as blowing up from below.

WHEN TO USE PLANT FOOD

If you're using a compressed or tablet form of

plant food, two or three tablets (or the approximate equivalent of whatever fertilizer you prefer) can be put into the soil within the first week or so after setting the seed into the pot. Within the next few months another like amount can be added, and so on.

There's no way of setting up a fertilizing schedule. If you think the plant's leaves are looking a bit bedraggled, or yellowing too often, add a booster to the soil. One can overfeed a plant, but I follow the principle that if the plant is growing well, and it is more than three months since plant food was added, it's time to push two or three tablets into the soil again.

A gardening note. After a while the surface of the soil becomes packed rather solidly in the pot. When this happens take a fork and gently loosen the dirt. Turn it over, digging not more than two or three inches below the surface. This serves to aerate the soil, and is beneficial for your plant.

5 ❧ CUTTING BACK ON THE YOUNG PLANT

❧ *THE YOUNG PLANT*

You have already cut back on your plant. The main stem was shortened while the seed was still in water. No more cutting back is required for a young plant.

About the time the main stem puts out its first shoots of leaves, a secondary stem may have begun to emerge from the base of the seed (or it may have been visible from the very beginning). Don't cut back on that secondary stem. It can grow up and become an adjunct to the main stem, putting out its own leaves, and making a double-trunked plant. A seed may show more than one of these tiny, tentative stems, and at times even a third or fourth may be seen. Leave them alone. The second stem will grow or it won't, while the remaining stalks often don't come to much. Wait and watch.

☙ *THE MATURE PLANT*

The avocado is a tree. In your house plant the main stem will become the trunk of that tree. From that parent trunk stems will grow, and will in their turn become branches. And from those branches still other stems will grow. And all of these stems and branches will be leafed.

Cutting back keeps a plant's shape within attractive and practical bounds. But cutting back also serves to stimulate the growth of dormant stems and branches that are prevented from developing because of too vigorous and rapid *outward* and *upward* growth.

The young branches that grow out from the *bottom* of the parent trunk should be encouraged. To do this, cut off stems that grow on the next *upper* set of branches. Go on in this fashion, always working down from the top of the plant. Cut off all younger stems and shoots of leaves that give a lopsided look to the plant as a whole. Some of these stragglers are temptingly green. But they have to go if you want your plant to be fully stemmed and leafed, with a symmetrical, *contained* shape.

Don't be afraid of cutting back. You can't

WHERE TO CUT BACK ON
THE MATURING PLANT

do the plant any serious damage by removing a crown of leaves or even whole and fully leafed stems. The avocado is tough and grows rapidly. Be bold; experiment with it. And watch. Successful growers all started by being good watchers.

Use a scissors. All branch cuts should be

THE HEALTHY PLANT, CUT

BACK FOR SYMMETRY

clean ones. Cut close to the trunk and branch so no unsightly stubs are left. You can pinch off single leaves and crowns with your fingers. Pinching is merely another form of cutting back.

Remember to trim from the top and the outside. That way you will force the plant's growth down, into, and toward the already existing greenery.

❦ THE TREE

Continue to apply the above.

The avocado has a resilient branch that dips and curves down as it grows out and away from the trunk. Sometimes one of these branches will reach out until it is nothing more than a bare stretch of leaf-tipped stem. If you cut off the whole thing you will soon see how the other branches benefit from the removal of such an unproductive member. New stems with budded leaves will appear in unexpected places, on lower, older branches as well as on other stems and branches.

Another gardening note. Make a practice of spreading pinched-off or dropped leaves across the top of the soil. This helps hold down evaporation of surface moisture and is especially useful in the dry winter months. As they decay, the leaves may nourish the soil.

THE TREE TRANSPLANTED
TO A WOODEN TUB

6 ❦ TRANSPLANTING INTO A LARGER POT OR TUB

If you have the success with your avocado that you can expect, you may soon begin to wonder when to transplant it to a larger pot or tub.

The time to do this can be decided by a sight judgment. If your plant *looks* too large for its pot, it probably *is*. A plant that is six feet tall in an 8½-inch pot needs to be moved fairly soon. If your plant is in a 10½-inch pot it can wait until it is well-branched and tall, but its appearance, in relation to the diminishing size of the base, soon begins to suggest that the time has come for a move.

I'm not quite certain what is the maximum size of terra-cotta pots, for I've seen some very large ones. But as their size increases, they also

47

become heavier and more difficult to manage when filled with earth.

A good-looking, octagonal tub is available on the market. Made of redwood, it is reinforced with horizontal brass bands. It is eighteen inches in diameter at the top, tapering down to a fourteen-inch base, and is fifteen inches high. The redwood tub usually has a good, generous hole in its bottom. Any variation of this size is satisfactory; just be certain that there is a good-sized hole in the bottom of whatever container you use.

Finding a dish to go under the tub may present difficulties. If you can't find an aluminum dish, try to get one of the hard rubber ones that are sold in gardening supply stores. They have a set of discs at the bottom that keep the plate from sticking to the floor. You need to have a plate that can be turned without too much trouble so that the plant can be moved in the light from time to time. (I keep a scattering of small pebbles in the dish under my largest tree.) Or, if you can find one, use the large-sized terra-cotta dish. You can put it on top of a rubberized mat, cut to size, and your floors or rug will be protected against seepage.

Prepare the base of the tub with a good supply of broken shards. You can use parts of the pot in which the plant is held, for you will have to break the pot anyway to get your plant out of it. Do this by first setting the plant on newspaper. Then, using a hammer, smash the pot with the plant in it. Break up shards and arrange them across the hole in the tub. Put soil over the shards and make a generous space for the plant to comfortably fit inside. Lift the whole mass of roots and as much soil as will come along, and move the lot into the new vessel. Shove the dowel into place, right down to the bottom of the tub. And then fill the tub with soil. Don't pack down the soil, and don't get the dowel too far from the trunk of the tree. Tie the dowel to the trunk, in the places that will best help support the branched tree.

Watch the tree flourish in its large, new growing space.

LIGHT FOR THE TREE

The tree should have lots of light. If you don't have sunlight, try to arrange for some sort of artificial light that will fall on the leaves from top to bottom for as many hours as can be managed. The more light you give the tree, the better it will grow. A floodlight would be ideal.

Turn the tree around from time to time. The shape of your tree's growth and development is affected by the directions from which light reaches it.

⟨⟩ WHEN.TO USE THE STICK OR DOWEL AS SUPPORT FOR THE PLANT AND TREE

The time to use the dowel becomes obvious early in the avocado's growth. The support that the plant and the tree get from the stick is important to its development. All sagging and heavily leafed branches should be tied to the dowel. Don't tie too tightly, or too closely. The stems can be injured by the weight of branches pulling against the trunk.

⟨⟩ PREFERABLE TIME TO GROW OR START PLANTS

An impression I once held was that the avocado grew better if started in the spring of the year. I've long since given up the idea. The plant can be started, and will grow well, at any time of the year. It only needs to be kept warm, moist and protected against extreme cold.

⟨⟩ RESTING PERIOD OF THE AVOCADO

All plants have a resting period. It's usually in the winter. (In the case of the avocado, the resting period is the dry season.) I'm not quite sure when the avocado's resting period really occurs.

50

I've seen the plant's growth slack off in the winter as well as in the summer. And each time I've said, knowledgably, here's the resting period. But I'm not at all certain that such was the fact. However, when there is a slowing down of growth, no signs of new shoots or leaves for a while, followed by a sudden spurt of activity, the pause between may well have been your plant's resting period.

There's nothing to do but leave the plant to its respite and watch for signs of new growth. And meanwhile keep it under the same conditions as before, warm, wet and well lit. Drop a bit of food into the soil—it helps the gardener, and might be useful to the plant.

WASHING THE LEAVES

The avocado, which is, after all, an outdoor tree, has a fairly tough-textured leaf that does not require protection against accumulated dirt and grit. Because the leaves look best when they are shining clean, the job becomes a housekeeping chore rather than a gardening one.

Wash the leaves with a wet, very soft cloth, and a controlled attack. The leaves will crush and become cracked if your washing is too vigorous. Be careful, and use tepid water. Don't use soap.

51

VARIETIES OF GROUND COVER:

1. GRASS 2. CHERRY PLUM TOMATO

3. UNIDENTIFIED BUT PRETTY WEED

4. BASIL 5. GRAPEFRUIT 6. LETTUCE

7 ❧ GROUND COVER

The outdoor avocado tree needs ground cover. Ground cover keeps the soil around the tree's base covered and moist, produces fertilizing material for promoting growth, and can be very attractive. An indoor avocado, however, needs nothing but water, food, light and attention. It is the gardener who will enjoy planting the soil at the base of his indoor tree with small green crops.

The possibilities for gardening the soil at the base of your avocado tree are almost endless. Complementary foliage gives a handsome effect to the tree's own greenery.

Another, and incidental, gardening note: If the soil at the base of your tree is covered with pinched-off leaves, turn them over and work them into the earth, using a fork. This is good general practice, and develops a mulch that is helpful to the tree itself as well as to whatever small plants you may grow.

I've grown grass seed in the middle of the winter. Sometimes it takes, and sometimes it doesn't, but it's fun to try. You can also plant grapefruit pits. Make small holes with your finger, drop in the pits, and then cover them. You can also plant pits from mandarin or temple oranges this way. Grapefruit and orange pits sometimes become charming seedlings, sprouting up into tiny green plants. But don't try anything that has too deep a root system, such as sunflower seeds and their like.

Try for small, young, shallow-rooted plants.

Last spring I planted three cherry plum tomato plants with two basil plants at the base of one of my indoor avocado trees. Both tomato and basil were bought as seedlings (young plants) from a gardening store that supplies summer truck gardeners. The basil can be started from seed, and it grows well.

I also planted beans, putting them into the patch of soil at the base of another avocado plant. I used dried red kidney beans washed in water and then permitted to sprout on a ground of wet cloth in a shallow dish. (You can have bean sprouts all year round this way, using dried kidney, lima and pinto beans.)

Within a short time we harvested a very small but astonishingly successful crop of tiny red cherry plum tomatoes, baby green beans and fragrant basil. All were delicious and gratifying to eat, and delightful to watch growing in our living room. We dried the basil leaves and stored them for future use.

You can try anything that will do well in the avocado's soil: radishes, young lettuce, tiny carrots, chive—the makings of a home-grown salad. Combined with a ripe avocado, they make a very tasty salad indeed.

Watch out for the grapefruit and the oranges—you don't want to start a contest between two trees. The citrus fruits grow alongside the avocado in Florida, California and points south. There's no reason why they could not become companion trees indoors, as well, especially if you are giving your avocado tree all the light, water and warmth it needs. And, now and then, don't forget to add a dollop of plant food.

I wish you good growing, and good eating.

A FRUIT-BEARING, FULL-GROWN
TREE, OUTDOORS

8 ✾ THE BOTANY OF THE AVOCADO

Avocado, of the family Lauraceae, genus *Persea*, species *americana*.

✾ FAMILY

The avocado belongs to the Laurel family, Lauraceae. The family is made up of tropical and subtropical trees and shrubs including cinnamon, cassia and the avocado, as well as numerous medicinal plants. In the United States the family Lauraceae is represented by sassafras, spice bush and Oregon myrtle.

✾ GENUS

The avocado tree is variable in size, from short (fifteen feet) and spreading to tall (forty feet) and slender. The wood is light and brittle, the leaves commonly elliptic or oval, from four to sixteen inches in length. The tree is an evergreen, its leaves growing all year round, although leaves are lost when the tree flowers.

Flowers are in dense racemes (the lilac's flower grows in racemes), about ½ inch across, more or less covered with hair, pale green-yellowish and lacking petals. The flower has a one-celled ovary. It is a monoclinous plant, which is a group having staminate (male) and pistillate (female) flowers on the same individual tree, a hermaphrodite. The single seed has two large cotyledons, and is tight to loose in the cavity.

The cotyledon, which is also called the seed leaf and is the first pair of whorls of leaves in the embryo, serves to supply stored food to the plant until it is mature enough to begin photosynthesis. Photosynthesis is the process by which a plant makes its own food, using sunlight, from the atmosphere and from water and minerals in the soil.

❧ SPECIES

The avocado, which is indigenous to the New World, was originally divided, horticulturally, into three different groups or races: the Mexican, the Guatemalan and the West Indian or South American.

Later botanists, however, have since claimed the Mexican tree as a distinct botanical variety, *P. americana drymifolia.* This tree is native to Mexico. Its leaves are quite small and are anise-scented. The fruit is small and round, from three to eight ounces. The skin is smooth, light green; the seed coat is thin; the cotyledons comparatively large and smooth, tight in the cavity. This is the hardiest and most frost-resistant variety. It served as the root stock for the later commercial California avocado.

The Guatemalan variety is native to the highlands of Central America. Its leaves are larger than the Mexican, and its fruit is variable in size, from ten to thirty-five ounces, and pear-shaped to flattened oval. The seed coat is nearly smooth; the cotyledons almost smooth, rather tight in cavity. It is slightly less frost-resistant than the Mexican.

The West Indian or South American variety is the most tropical. The leaves are very large (to sixteen inches) the fruit is large (up to three pounds) and pear-shaped to oval. The skin is thick, leathery, dark purple in color, and pliable. The seed coat is rough-surfaced and thick, the cotyledons larger than the others, usually separated and loose in cavity. This is the least frost-resistant variety. Today's commercial Florida avocado is derived mainly from hybridized Guatemalan and West Indian varieties.

The fruit of the Guatemalan variety takes nine to twelve months to ripen on the tree, the West Indian six to nine months, and the Mexican somewhat less.

As a result of extensive commercial hybridization among all three varieties, great changes have been brought about in fruit size and shape, as well as in other characteristics. It is doubtful whether any pure Mexican, Guatemalan or West Indian trees exist commercially in the United States.

POLLINATION

In expectation of the question you will be asked when your avocado seed has grown into a house plant—"Will it bear fruit?"—the following may be useful.

A single avocado tree can and does produce fruit, for the avocado has a perfect flower with both staminate—male—and pistillate—female—parts. The stamen contains the male gametes, the pollen; the pistil produces the female gamete, the egg. The avocado flower has twelve stamens, each with four pollen chambers, and a single pistil with one ovule.*

* To underline the old saw, there is an exception to this rule, too. For in some few varieties of the avocado, the flower produces no pollen at all.

The stigma is the apical portion of the pistil that receives the pollen, but in the avocado the stigma of the flower often matures before the pollen. It is therefore unlikely that self-pollination will occur within a single avocado flower.

The pale-green flower grows in dense clusters, called racemes. Each raceme is supported by a stalk, branching from a single axis, and is composed of small clusters of flowers, each individual flower, with its pollen and ovule, developing as a unit.

If pollination between two flowers is to be effected, the two flowers must be in the same phase of mature development. The time lapse between the maturation of the stigma and the stamen must be by-passed or over-lapped in order for pollination to occur between two flowers on a single tree. Fertilization can be effected on a single tree where one flower containing a mature stigma is pollinated by the mature stamen of another flower, but one that blossomed earlier. In other words, the separate flowers can catch up with each other's maturation.

The answer to the question whether an indoor avocado plant or tree will produce fruit is a qualified maybe but not probably. Unless you live in an enormous duplex apartment, can spend up to seven years cultivating the tree, and can comfortably accommodate a hive of bees, or know how to pollinate manually, your indoor tree is likely to remain fruitless.

However, the dooryard trees that are so popular in many parts of the tropical and subtropical world frequently bear fruit, and theoretically, so can an indoor avocado tree.

I've never had flowers on any of my trees, let alone fruit. But perhaps someone will be heard from who has grown an indoor avocado or two, and if they have, I'd very much like to hear about it.